Made in United States
Cleveland, OH
04 November 2026

47647248R00069